Travis Dee Bechtol

Date: 10-17-2024

Science and it's magnetic pulsation, throughout the seasons

- As the warp is around all of Earth and everywhere out in the Universe. It is creating a pulsation down on all of us. Which in return makes a dynamic motion throughout the seasons.
- The seasons changing every year make me wonder, what word and such, is making the temperatures change.
- You can see in the pictures below, that the seasons changing around, are making a big magnetic pulse of pressure, at this location here on Earth.

The universe in motion, never ending change

- Arguments and different perspectives come down too, many things as in God, to also the creation like the Big Bang theory. I do not believe in the Big bang theory, like a lot of others.

- I more on the side of what people are thinking about, that is, we were always here. This universe was always here, it never was created. It was always the same, it never moved, never changed.

- I'm on more proposals of people who are saying, that this was changed over and over on Earth, from the people here. That it was always already in stable and unstable form.

Title of a concept in matter

- Earth and the Universe are an interesting outlook, over everything else out in the sites. Waves and storms are nice to see and look as they come by every year. Clouds and the static, with atoms, particles, the things you cannot see. Showing us that we do not control but have control.
- The moon causing it's bound back on Earth is interesting. Looking back how Einstein and so on, would later prove with mathematic formulas. That the moon and the earth lined up in a solar eclipse making a long causes a perticritical wend, back on out showing an interesting outbound.
- Which would later keep being analyzed by many and not for sure, just all the credit to one person. Many people were and are a part of the outcomes.

Winterizing the change from, no winter or winter. Making the differential sequence

- North America at our location in the U.S.A.. We have a very good position in some sorts. Yet to me the argument is that we are getting rid of our whole position.

- If we keep getting rid of our land, we won't have the minerals from this top region at its level. Covering it with houses, neighborhoods and roads, are terrible for the future.

- The pulse that is coming from below and above back on to it, makes everyone so much sicker, with illnesses, along with losing brilliance.

- The Wi-Fi manipulation, we have here now in the world. According to that machine, is a method that can easily be pulsed by other machines, creating no correct metaphor.

Light and it's change over the greenage and Northern belt

- This image is from a different countries satellite. It shows that their region is much higher towards the North of Earth than what we, over in North America are, in the last pictures slide.

- In this image it shows great vegetation over on our old land, we came from. It seems they are doing a great thing there. Taking down the cities and metroplex business. And looking more forward to the science part to survive.

- Makes me wonder if the country were in, here in North America, are being built by someone, that is not from our own.

Colors and temperature changes, cell boost, and the flow of all and humanity. The change, and the detriment of life's cycle.

- The sun and its light, with rays, and beams, as it flows in and out of a pulse, that takes on all of the forces back in and out of a verse.
  Making a out form over the rotation of the planets, earth, sun, and stars. The mountains and snow, push a over cell, which pushes.

- The colors of the blue, mixed with yellow, and the suns set, shows a out mix, to the clouds, and ground below.

1 2 3 4 5
a b c d e
9 3 6

Max
$\frac{1}{6}(1-3+4)$

A32X

S36

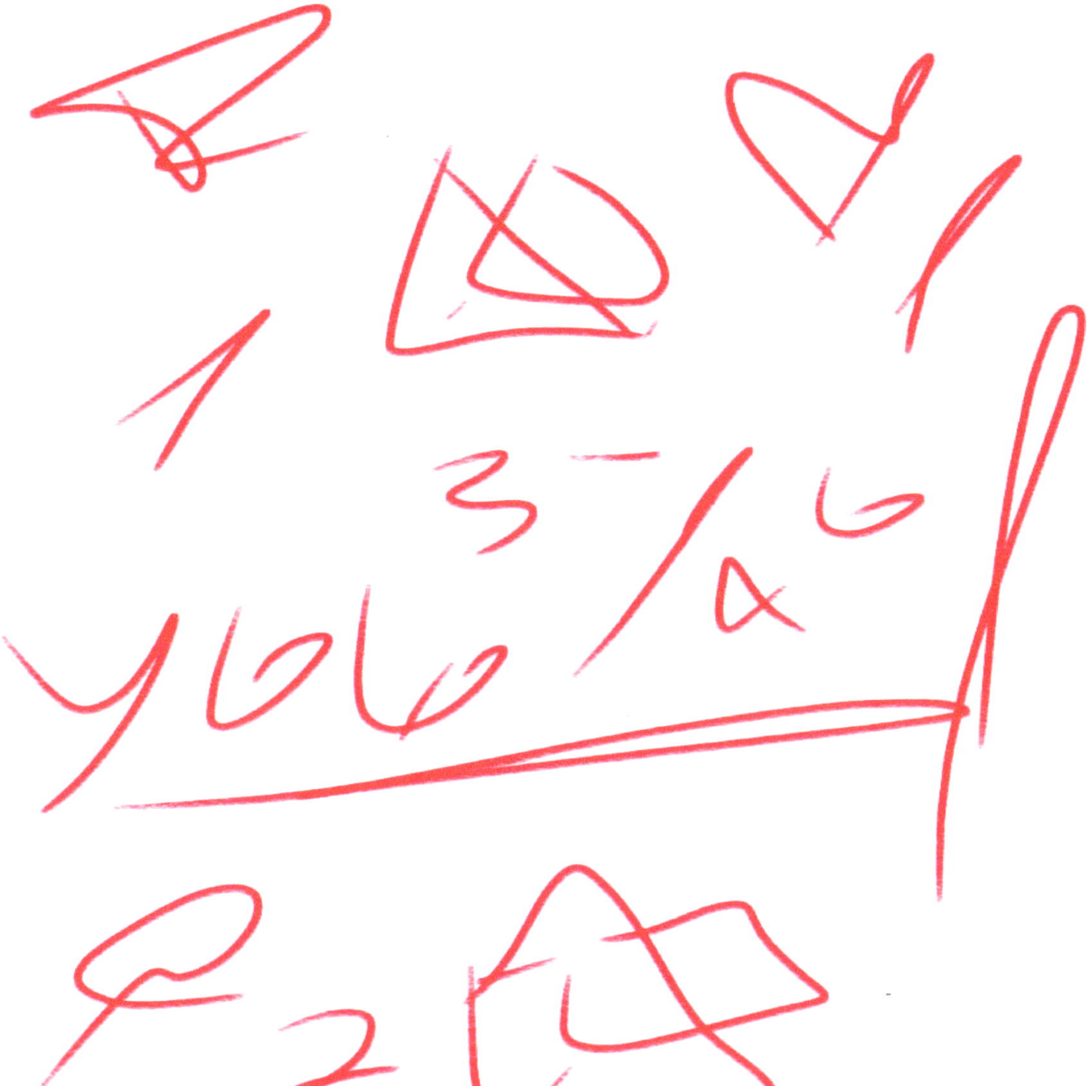

- The critical take on humans, and the world over forces back on us. Show that we can be inside of relativity. But learning in and outs, which is like a reduction formula when it always comes back to where you are at all times.

- But our machines and science we have a making a trigonometric spring, in Math terms. Convergence is always at our cycle; it never seems to surpass the same resolution.

- When will we go into the next level? Do we continue to add and write into it? Or do we go back some and see when times moved faster based off precise speaking outload communications.

- Like a differential equation, maybe breaking through the circle in it. Can bring us into examples that push further past what we see. As the writing continues, we will see a change in the dirt, to the water, onto the growth of vegetation inside us.

- Maybe we can accelerate into a direction you believe. It is writing, it is math, letters, symbols, and numbers; it creates a recipe, that you eat, and you grow with continuously till you die, or write on?

The End

Published on Amazon Kindle Direct Publishing

Mathematical drawings made on, Sketchbook application at Google Play Store

Thank you for reading,

A Pleasure,

Travis Dee Bechtol